Einstein's Cosmic ether, the Atomic ether, their etherons; small Big Bangs, gravity, light and our mind

*Science should never refuse to study **all** phenomena in the universe, from material to immaterial, from natural to so-called "supra or para"-natural.*

by

Prof. dr. Egbert Duursma[1]

and

**Prof. Dr. Ioan-Iovitz Popescu
Inventor of the Etherons
(Foreword)**

---

[1] Member Academia Europaea, former director Netherlands Institute of Sea Research and of the Delta Institute of Hydrobiology of the Royal Netherlands Academy of Sciences. e-mail: duursma@orange.fr

This document is subject to copyright. All rights reserved, whether the whole or part of the material and subject concerned, specifically the rights of translation, reprinting, reciting, broadcasting, reproduction on Internet and production of TV series and films.

© September 2017

2nd edition (With additional French summary)

Cover Photo: author

**Foreword to the second edition
by
Prof. dr. Ioan-Iovitz Popescue[2]**

Congratulations for your breakthrough trial of the old paradigm of the subtlest state of matter, supposedly filling the whole of our Observable Universe, the Ether, made of about the 10^{122} tiniest conceivable particles, The Etherons.

Particularly, you dared for the first time to forward the strong hypothesis of Etherons explosions, as SMALL BIG BANGS, happening any time and triggering chain reactions leading to all known building blocks of Nature, from elementary particles up to the stars.

Note that an electron should be a quite stable vortex made of about 10^{39} etherons, and a proton of about 10^{42} etherons. By this daring hypothesis of Etheron Bangs, you open the way of understanding the gap between the subtle Ether matter and the common atomic-molecular matter. This represents, surely, a big step in human knowledge.

---

[2] Em Professor in physics, and discoverer of the etherons in 1982. See therefore Duursma, E.K. editor 2016. *Etherons as predicted by Ioan-Iovitz Popescu in 1982*, specialist in Plasma Physics., Bucharest, Romania

Why this document?

*Cosmic and atomic small particle research absorbs billions of dollars, while that of **the ether of the universe and of the atoms** is practically neglected. However, these atomic spheres and the universe contain **etherons** of a size of 10^{-35} m and in a quantity of dozens of orders of magnitude larger than that of the particles of atomic nuclei of matter. They play a basic role in nature and life. Mankind's spirit, bound to the atomic ether, has not reached yet its optimal level, allowing all kind of failures and excesses.*[3] *Then what is this atomic ether, which seems to be so imperative?*

---

[3] **Read also**: E. K. Duursma, 2012. The Zwikker Code; Clearing mankind's indoctrination, Amazon Kindle e-book.

CONTENT
- Cosmic ether
- Atomic ether
- Creation of atomic ether
- Etherons, properties and Small Big Bangs
- Major genesis and perseverance of the universe
- Mind and atomic ether
- Characteristics of mind
- Matter and mind
- Recommendations
- Conclusions
- Summary & Résumé

Cosmic ether

Albert Einstein, in an address delivered on May 5th, 1920, at the University of Leyden: *According to the general theory of relativity, space is endowed with physical qualities; in this sense, therefore, there exists an **ether**.*  *[additionally,] space without ether is unthinkable; for in such space there would not only be no propagation of light, but also no possibility of existence for standards of space and time (measuring-rods and clocks), nor therefore any space-time intervals in the physical sense. But this ether may not be thought of as endowed with the quality characteristic of ponderable media, as consisting of parts which may be tracked through time. The idea of motion may not be applied to it. The ether does have electromagnetic properties (permeability and permittivity), from which Maxwell deduced the speed of light.*

Since the Hubble telescope discovered that our universe is expanding with an increasing speed, explanations have been sought by a kind of energy which causes this antigravitation phenomenon, a dark energy. It is also called zero energy which some inventors seem to be able to use for producing mechanical energy.

The leading view is that the universe since 7.5 billion years after the big bang is expanding with an accelerating rate, because "dark energy" is counteracting gravitation. So far nobody knows what dark or zero energy is, and it is **very strange** that such an energy whose dimensions are unknown, is **"pushing"**. All this should be related to the cosmic ether mentioned by Einstein and it becomes obvious that more knowledge must be gained about the emptiness of our universe, which is not only limited to space in the Universe, but it is also part of all matter. Therefore a chemical perspective on the ether phenomenon ether can add some details to this discussion.

Atomic ether

In 1913 Niels **Bohr** published a theory about the structure of the atom, based on an earlier theory of Ernest **Rutherford**. This last scientist had shown that the atom consisted of a positively charged nucleus, with negatively charged electrons in orbit around it (**Fig. 1A**). Bohr expanded this theory by proposing that electrons travel only in certain successively larger orbits. He suggested that the

outer orbits (or shells) hold more electrons than the inner ones, and that these outer orbits determine the atom's chemical properties (**Fig. 1B**). The electrons are present in an empty volume a thousand-billion times larger than that of the nucleus of atoms, while their proper electron volumes are negligible.

Fig. 1A.
- An atom contains a nucleus of protons and neutrons which have a radius of: 1.6 (Helium) to 14 (Uranium) fm. (1 fm=10^{-15} m)
- The radii of the atoms of Helium and Uranium are 60 and 255 pm (1 pm =10^{-12} m), respectively.
- From this, the ratio between the volume of the atomic ether to that of the nucleus is: for Helium: 54×10^{12} and for Uranium: 6.0×10^{12}, (10^{12} is thousand-billion).

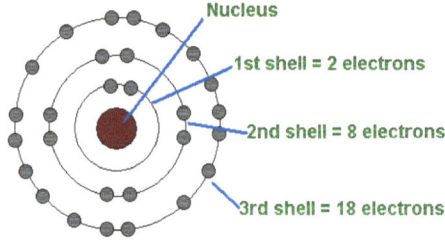

Fig. 1B.
The distribution of the electrons in the atomic ether is in shells, of which the outer shells contribute to the formation of molecules.

- All matter in the universe, **except** that in neutron stars and black holes and that of cosmic ray particles (90% protons and 9% alpha particles) consist of atoms or molecules of the elements of the periodic system, and is for more than 99.999% empty. This emptiness we denominate as the **atomic ether**.
- This atomic ether is uncompressible as it seems, and has similar properties as the cosmic ether in permeability and permittivity for some electromagnetic radiation and gravity.

- Although difficult to estimate **(Fig. 2)**, cosmic ether would be equally uncompressible since having the same properties as the atomic ether, except the latter containing electrons in different orbits. The question is, however, by **what** can it be compressed? Nuclei of material molecules are infinitely smaller.

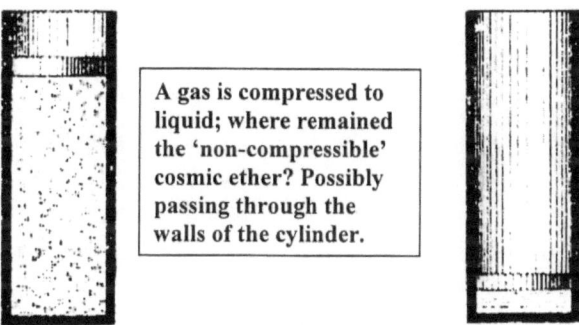

A gas is compressed to liquid; where remained the 'non-compressible' cosmic ether? Possibly passing through the walls of the cylinder.

Fig. 2. Imaginary experiment of compressing gas molecules, free moving in cosmic ether.

Creation of the atomic ether
- If we take the alpha particles, (consisting of 2 protons and 2 neutrons[4]), which arrive on earth from cosmic rays and those which are produced from radioactive substances, in particular the transuranic elements such as Uranium and

---

[4] An unsolved enigma is why only the combination of 2 protons and 2 neutrons are released from radioactive substances and not any other combination of protons and neutrons. It has probably to do with the basic inter-nuclei forces. For heavy particle radioactivity see more at https://en.wikipedia.org/wiki/Cluster_decay .

Plutonium, the question arises **how** is the atomic ether of the produced Helium **created**?

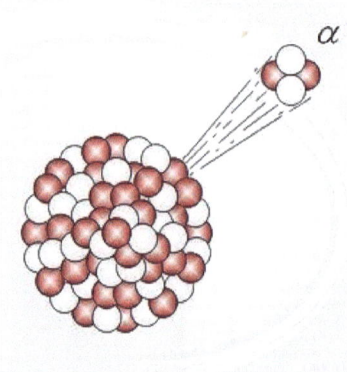

Fig. 3. An alpha particle (2 protons + 2 neutrons) emitted from a transuranium nucleus.

- If positive alpha particles would first capture 2 negative electrons from its environment (**Fig. 3**), than the alpha particle would become 4 neutrons. However, neutrons are instable and within minutes they convert into protons and β particles. These β particles (electrons) should be projected with great speed as secondary radiation. Such a process has never been observed from α radiation in a magnetic field (**Fig. 4**), where the α particles are going to the left, and the β particles to the right.

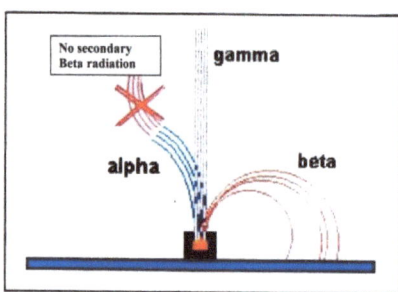

Fig. 4. Alpha (2+) and beta (1-) particles undergo deflection in a magnetic field. If alpha particles would absorb electrons, this would convert protons into neutrons, which are unstable and would reconvert again in protons by ejecting beta radiation. **This is not the case.**

Thus, there remain three possibilities, where
- the alpha particle **first** captures its atomic ether sphere of thousand billion times its volume, and then absorbs 2 electrons from its environment, or
- the **other way around**, or
- **both** at the same time.

Since cosmic ray alpha particles **do not** capture anything from the cosmic ether, it seems logical that the second possibility is most obvious, although since it happens extremely rapidly, it approaches the third possibility. When cosmic alpha particles hit the atmosphere, they are converted into Helium gas by "grabbing" electrons and atomic ether from existing gas atoms or molecules. This should be the same for alpha particles from radioactive substances, hitting material or gas molecules.

The bulk of the neutron stars is next to neutrons, protons and electrons.[5] They have about 1.4 solar masses, but only have a diameter of 12 km. Remembering that neutrons have in principle a short life of tenth of minutes, and that any element with a number of protons + neutrons in the nucleus higher than about 250 is instable, the question is what kind of processes occur inside of the neutron star. Any neutron in the core can become a proton which emits an electron, but this electron can be recaptured by another proton within the core. The surface of the star will be made of hydrogen and it is source of consequent

---
[5] https://en.wikipedia.org/wiki/Neutron_star

radiation from radio waves to x-rays and gamma rays. The amount of neutron stars in the Milky Way is estimated at several hundred million.

Are they all, including the larger black holes potential sources of "Small Bangs"? It would be interesting to know. If so, the produced nuclei will "grab" from the cosmic ether the required atomic ether in order to create atoms and molecules. Probably sufficient electrons are around at such "Bangs".

Etherons, properties and Small Big Bangs

For any electromagnetic transfer in the universe and within atoms, a medium is requested that passes on radiation, but also "handles" the forces like magnetic forces and gravitation. From one spot to another, this medium should pass on these forces, like iron particles do with an iron magnet for magnetism (see **Fig. 5**).

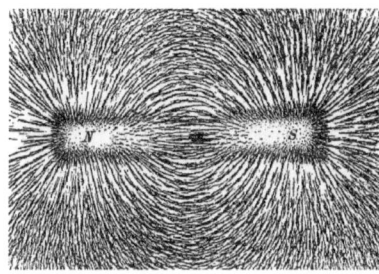

Fig. 5. Iron particles in a magnetic field.

This brings us to the theory of the existence of hypothetical particles (units), which the cosmic and atomic ethers contain. Ioan-Iovitz Popescu[6] gave them the name of **etherons**. It could have been gravitons, as it has been some-

times suggested, but the essence is the part of the word "ons", which has been used so fre-

---

[6] Popescu, I.-I. and Nistor R.E., 2005. Sub-quantum medium and fundamental particles. Rom. Rep. Phys., 57, No. 4, 659-670., see at
https://arxiv.org/ftp/arxiv/papers/0908/0908.1945.pdf

quently for so many sub-micro particles like protons, geons, neutrons etc.[7]

Popescu suggested them to have a rest mass of 10^{-69} kg with a rest energy of 10^{-33} eV/c^2, and a radius of about 10^{-35} m, existing each at a distance of each other of 10^{-15} m, which is the size of an atomic nucleus (See also **Fig. 1A**). This would result in $2.2 \times 10^{+14}$ etherons per atom of for example Helium.

Much larger is the amount of etherons in the (observable) Universe, namely 10^{122}. It was estimated by Popescu[2,6] and as number later discussed by Funkhouser[8], taking into account the still unknown extension of the Universe. Thus basically the etherons carry $10^{122} \times 10^{-69}$ kg = 10^{53} kg of the mass of the Universe out of which the stars and planets carry about 10^{52} kg.[7,9] By contrast the volume of the Universe with respect to that of the etherons is void by a factor of $6.8 \times 10^{78} m^3 / 4 \times 10^{17} m^3 = 1.7 \times 10^{61}$ (**Table 1**).

---

[7] The list is very extensive, like Photon, Gluon, W-boson, Z-boson, Higgsboson, Graviton, Kaon, Pion, Meson, Lepton, Neutrino, Electron, Positron, Muon, Proton, Neutron, Baryon, Fermion.

[8] S.Funkhouser, 2008. A new large-number coincidence and a scaling for the cosmological constant. Proc. R. SOC. A. 464, 1345-1353, see also at:
http://arxiv.org/ftp/physics/papers/0611/0611115.pdf

[9] http://www.physicsoftheuniverse.com/numbers.html

Table 1. Data on etherons and the Universe.

Etherons

Diameter etheron	10^{-35} m	Popescu
Distance between etherons	10^{-15} m	Popescu
Volume etheron	4×10^{-105} m^3	
Weight etheron	10^{-69} kg	Popescu
Specific weight etheron	2.5×10^{35} kg/m^3	
Amount etherons in Universe	10^{122}	Popescu
Weight etherons in Universe	10^{53} kg	
Volume etherons in Universe	4×10^{17} m^3	

Universe

Radius Universe	1.7×10^{26} m	National Solar Observatory
Volume Universe	6.8×10^{78} m^3	National Solar Observatory
Ratio volume Universe/etherons	1.7×10^{61}	
Weight visible matter in Universe mostly hydrogen	6×10^{51} to 6×10^{52} kg (0.74 of all matter in Universe is hydrogen)	National Solar Observatory

Supposing that these etherons exist in either form of particles, mass or energy units, and suppose also that all matter/energy of primary particles of the Universe are of the **same nature** (a kind of basic substance), the etherons appear to be much more compressed than electrons and protons (see **Table 2**).

Table 2.

Particle	Volume m^3	Weight kg	eV/c^2	Density kg/m^3
Etheron	½x10^{-105}	10^{-69}	5x10^{-34}	2x10^{36}
Electron	½x10^{-54}	9.1x10^{-31}	5x10^5	1.8x10^{24}
Proton	½x10^{-45}	1.66x10^{-27}	9.4x10^8	3.3x10^{18}

The question can be posed whether the etherons are **unstable** with a kind of half-life of considerable time. Thus, when one etheron is 'bursting', it may give an impulse to a chain reaction of other bursting's and 10^{42} etherons produce one proton. Probably this occurs at extremely high temperature which can initiate the next process of nuclear fission of hydrogen

And here we touch upon what has previously already been suggested by Fred Hoyle([10]) of **Small Big Bangs** of his **Steady-State-Theory**.

Proof of such a theory is that there should exist stars being younger and older than 13.8 billion years ago, the time the BIG BANG was supposed to have occurred.

The team that used the Hubble Space Telescope and the ESO Very Large Telescope for taking images of galaxy NGC 4365, were able to

---

[10] Marek A. ABRAMOWICZ, « **HOYLE** FRED - (1915-2001) », *Encyclopædia Universalis* [en ligne], consulted on 30 July 2017.
URL: http://www.universalis.fr/encyclopedie/fred-hoyle/

identify star clusters that were only a few billion years old, while the majority were over 12 billion years old.[11]

The most distant and oldest object known in the universe is the Galaxy GN-z11 of 13.4 billion years in the past, just 400 million years after the Big Bang.[12]

Thus, we observe that any time later, small Big Bangs may have occurred, producing new stars. Their source, as for the Big Bang, was the stock of 10^{122} etherons of which at certain points etherons exploded, initiating a vortex of extremely high temperature in which protons were produced. These continued to fuse into stars, such as existing at present.

This simplistic view is intermediate of the two existing Big Bang and Steady-State theories, but having the advantage that the source of the material in the Universe, the etherons, is established.

Properties

Suppose that these etherons really exist in either form of particle, with mass or energy, unit or another, **what should be their properties**?

- They should have electromagnetic properties in order to pass on: Radiation of any form and Electrical charge. In that case the distance

---

[11] https://www.universetoday.com/8416/young-stars-in-an-old-galaxy/ . Article written: 27 Jun , 2002. Updated: 10 Jan , 2013 by Fraser Cain

[12] http://www.dailymail.co.uk/sciencetech/article-3475182/Galaxy-smashes-cosmic-distance-record-set-astronomers.html

between these etherons in the Universe should be **smaller** than the wavelengths of the various radiations. As shown in **Fig. 6.**, these wavelengths range from 10^{-15} m to 10^1 m, which is above the distance of 10^{-15} m between the etherons.

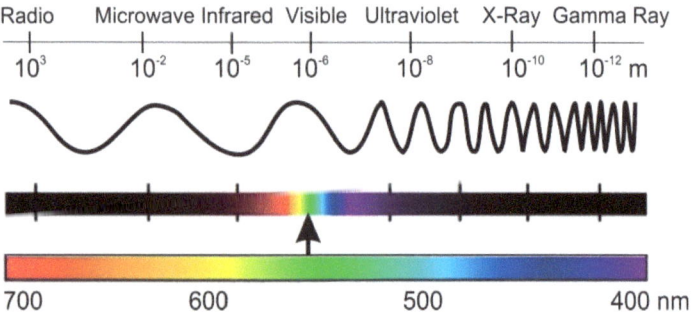

Fig. 6. Wavelengths of radiation « passed » by etherons. Note that these are all larger than the distance between the etherons of **10^{-15} m**.

- They should "pass on" Magnetism and **Gravity,** the latter in a similar way as shown in **Fig. 5** for iron particles in a magnetic field. In that way the gravity attraction of one mass by another mass becomes obvious.

Thus we finally know how it is possible that one mass "knows" it is attracted by another mass.

Are the etherons stagnant and in constant amount per unit space?

- **Within an Atom:** Since they are equally attracted by gravity, such as air molecules by the earth, the etheron concentrations in the atomic ether might be higher (or more compressed) closer to the nucleus, thus explaining perhaps the zona-  tion of electrons in the atomic ether as explained in **Fig. 1B**. This was the basis for the Periodic System, as developed by Dmitri Mendeleev (**Fig. 7**).

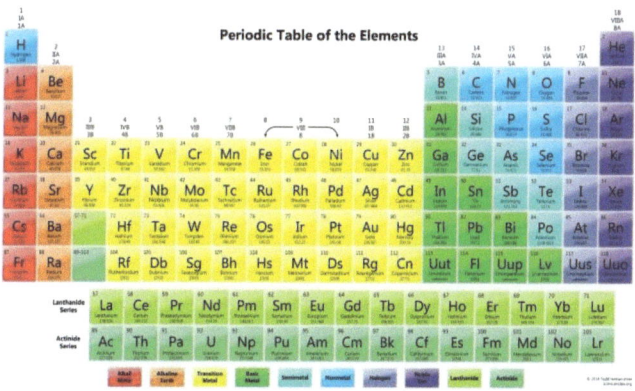

Fig. 7. Periodic system of elements, where the number of an element represents the amount of protons in the nucleus and the amount of electrons in the surrounding sphere.

- **In the Cosmos:** Since light is affected by the black holes, this would indicate that in the "neighbourhood" of extremely high centres of gravity, the concentrations of the etherons may also be higher (or more compressed) than elsewhere in the universe, thus bending the light beams coming from "adjacent" stars.

How would the "passing on" of radiation and gravity occur?
- For radiation, the etherons are in a state of vibration that allows the radiation to pass on with the speed of light. They don't move themselves but act as a kind of cilia.
- For spiritual activity in our brain or thoughts between individuals at distance, this is more difficult to explain, while the thoughts may flow independent of the speed of light, which means also more rapid.
- For gravity this is a state of stagnancy, existing rather than occurring.

Can transfer of radiation and gravity be blocked?
- For radiation a blockage may indeed occur, depending on the intermediate matter and kind of radiation.
- For gravity no physical or material blockage seems to be possible. Any object in between two masses plays a role of additional mass, even their own atomic etherons.
- However, if telekinesis caused by a human mind exists, it means that there is an interference with gravity attraction of objects by the earth.

Major genesis and perseverance of the universe
Accepting the theory of the atomic ether, the **real creation of the universe** started by the fact that in the created atoms the negative electrons were **not absorbed** by the positive nucleus. The electromagnetic "motion" in the atomic ether remained stable for billions of years. Due to this motion in a "vivid" emptiness we may speak also of a "living" atom. Only in dying stars and black holes the atomic ether is collapsing.

Ignoring for the moment that we don't have a clear explanation for the "living" atom, its emptiness is nevertheless the **corner stone** of the universe. The question is what is in motion and how? Classic views consider shells of electrons, moving around the centre; quantum mechanics view a statistical energy distribution. Some radiations can pass through, some not.

In the emptiness of the atoms occur all the phenomena, which have led to form combinations between atoms such as inorganic and organic molecules, and this started on earth some 5 billion years ago, if not earlier by chemical reactions. The basis of chemical transformation is the rearrangement of electrons in the chemical bonds between atoms.

But what happened when the first living material was formed? Some external signal, such as electrical sparks, changed organic compounds in the sense that the emptiness motions were affected and became reproducible. A slight aberration caused its reproduction, and resulted in the formation of the first living material. From that on, evolutionary processes started to continue and finally resulting in the living world of today.

However, was this initial mystery only happening billions of years ago? If the above theory is correct, there is no reason to believe why this process of formation of living material may not **have continued for ever**. Therefore it would be worthwhile to study species, which have only **"recently" been created**

Mind and atomic ether
If we play a musical instrument, our mind instructs the neurons in our brain to give an electrical signal to our fingers. How is that possible? Only if somewhere in the emptiness of the atoms of the neurons, a signal is given which results into that reaction.

And here we are in the field of the **bridge** between matter and mind. It will mean that our spirit is capable to act in the atomic emptiness on the electromagnetic motions and reactions.

But where is this mind?
Can it act also outside of the body, since the emptiness of space is not limited to only molecules? If our spirit is able to act in the emptiness of the atoms of our body molecules, why not it can act also anywhere in the ether of the uni-

verse. The same would then hold for spirits of deceased humans.

Characteristics of mind
Definitions of spirit as given in dictionaries:
- The non-physical part of a person, regarded as their true self and as capable of surviving death or separation; or
- The animating or vital principle held to give life to physical organisms; or
- The immaterial intelligent or sentient part of a person.

Has spirit electro-magnetic properties?
- Probably, since it is present in an ether which supports electromagnetic events.

Where is our spirit acting in our body?
- Obligatory acting on the molecules of our brain neurons. And thus having access to the volume of the atoms and is able to affect electromagnetic processes.

Is mind in our body acting in the dimensions volume and time?
- Yes, because it can act on different neurons in our brain at any moment desired.

Is spirit limited to mankind?
- Certainly not. Many organisms have the capacity of acting by making a decision to act.

To which quantity?
- It is limited to the brain capacity and the number of available neurons, which are the highest in mankind, at least for people without brain damage.

Is mind limited to a living body?
- Since the emptiness of our atoms and molecules is almost equal to that of the universe, there is only a barrier between them of electrons. Probably mind can in fact communicate outside a body (brain) as far as physical-chemical properties concern. If this occurs, there is nothing paranormal involved.

Are the minds of deceased persons surviving death of the body?
- Since mind is acting in the emptiness of atoms and molecules, it can also survive in the emptiness of the surrounding universe. At least why not. Therefore religions classify their saints as survived minds. From the time of our prehistoric ancestors, this is thought to be true.

Can religious phenomena such as believe, faith, praying, meditation, acts of saints be explained in physical-chemical terms?
- Yes, in the form of seeking contact with either spirits of deceased ancestors and saints. Whether there is a common spirit as suggested by Carl Gustav Jung, depends of its interpretation.

Matter and mind
It is as old as the world that mankind tries to have an impact on matter and also believes or is certain, that matter has an impact on mind. For most of them there is no para-normal involved. Beautiful colours can make us feel a better men, and men tries to express in art a kind a radiation.

But limiting us to only those phenomena of impacts between matter and mind, which involve the atomic ether and cosmic ether, where man uses the atomic ether of its body cell molecules, and in particular those of the brain neurones.
From this viewpoint, it will be possible to find the structure and *modus operandi* of for example
- Telekinesis, which concerns the relation mind (spirit) with gravity, both active in the atomic ether,
- Telepathy,
- Hypnosis, which has an impact on our brain action, and
- Signals between humans at distance, such as between twins, which seem to act timeless.

Which forces is encountering mind (spirit) in its atomic and cosmic ether? As Einstein already said, *the cosmic ether has electromagnetic properties such as permeability and permittivity for letting radiation go through, and this count also for gravity.* In the atomic ether of molecules, we have additionally the binding forces which keep the spheres together and in which electrons of the outer orbits play a dominant role. We mentioned already a kind of "vivid" emptiness with full of action. But also full of something what is not yet understood. Words like *dark energy* does not help us further, since both *dark* and *energy* may be irrelevant, at least so far we know.

Fig. 8. International space vessel ISS. Courtesy of ESA

Telekinesis
Telekinesis is the movement of objects at a distance, supposedly without being physically touched. It is a hugely controversial phenomenon, in spite of the availability of a certain body of evidence for the existence of this phenomenon. But nevertheless telekinesis is in principle possible, it only depends on the strength of spiritual forces in the atomic ether and existing gravity forces. But these might be too different in strength for most people. A solid answer can be delivered when telekinesis experiments are carried out at the space ship ISS **(Fig. 8).** It requires only an astronaut with mediatic gifts.

Telepathy.
Telepathy is the name assigned to vibrations that propagate through the counter world. Telepathy belongs to a category of phenomena related to telekinesis, but is independent of gravity and

concerns information obtained from atomic ether, exterior of the person. These may be not or weakly related with time. And here we touch the history of events connected to matter. Religion does recognize this phenomenon such as the veneration of relics and statues, while also actively spiritually loaded objects are created, such as Sacramental or Communion bread (hosts). Is it possible that objects and matter have a kind of memory of events, which are sensible for mankind? On the basis of atomic and cosmic ether and the liaison between them, it seems "technically" possible. But again as for telekinesis it depends on the magnitude of these ether-related phenomena.

Hypnosis
Hypnosis is the influence of human spirit on that of others, but also of some snakes on their prey by hypnosis is well known. Again here, this phenomenon is only possible when the borders between atomic ether of brains and cosmic ether are trespassed. Without this, hypnosis will not have any effect.

Signals between humans at distance, such as between twins.
This phenomenon is also well known and the question is how these signals can pass in the field of atomic and cosmic ether. These signals seem to be independent of time and are extremely rapid. Since no mass is involved, it can be more rapid than the speed of light.

Recommendations

It is obvious that the atomic ether is a terrain for basic future research of physics, chemistry and psychology, which can be executed at much lower costs than the actual particle and space research. Its results would also have an impact on the research of the cosmic ether and the functioning of the universe.

- It seems logical to start with studies on the permeability and permittivity with respect to ultra-high frequency to lower-frequency radiation on atomic ether of various elements, from noble gasses to instable transuranics.
- Possibly nuclear fission studies contain aspects that might give light to the atomic ether.
- The same holds for the studies in Caderache on the future fusion nuclear reactor ITER **(Fig. 9).**

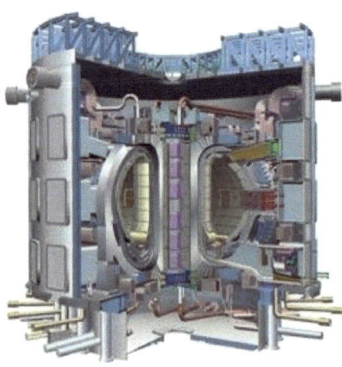

Fig. 9. ITER (International Thermonuclear Experimental Reactor) is a joint international research and development project that aims to demonstrate the scientific and technical feasibility of fusion power. Courtesy of ITER.

- Naturally the CERN **(Fig. 10)** in Geneva could focus on atomic etheron studies.

Fig. 10. CERN international centre for particle research at Geneva. Courtesy of CERN.

- Studies should be made on the working of our consciousness and sub-consciousness in how far these are impregnated in "our" atomic ether. Maybe we could solve the indoctrination of complete populations.
- Concerning the suggested small Big Bangs, it should be studied at which temperatures these could occur, in order to avoid that these are happening artificially.

Conclusions
- The **major mystery** of our universe is the fact that **atoms, containing large empty spheres (atomic ether), do not collapse for billion of years**.
- Both the Cosmic and Atomic ether contain etherons, of a size of 10^{-35} m. They can transfer radiation and are intermediate both for gravity and magnetism.
- The density of etherons, expressed in kg/m^3 is extremely high, which would allow them to explode and be the cause of **small Big Bangs**, which created young stars by hydrogen fusion.
- The etheron concentrations can be higher close to centres of high gravity, such as for atoms, near the nucleus and for the universe near black holes. That would explain zonation of electrons in atoms, and light bendig near black holes.
- The formation of atomic ether can be studied from radioactive α particles that convert into Helium atoms.
- If molecules can be formed chemically in the universe, the chance exists that some reactions may become reproducible and generate into living material. It may happen statistically or it requires an external impulse. This form of creation of life may still happen at present.
- The bridge between matter and mind should be located in the "not really empty" spheres (atomic ether) around the atomic nuclei and their etherons.
- The human spirit, alive or from deceased persons, may react in the emptiness of the universe with "help" of the etherons.

- As long as science has no more insight in the "behaviour" of spirit in the vast "vivid" emptiness of the atomic ether of our brain molecules, it will be difficult to evaluate many of the above described phenomena and theories.
- Science should, however, never refuse to study them and find reliable answers. There lays a large task for physicists and chemists in solving the permeability and permittivity of the atomic ether.
- A study by religions, accepting the existence of ether, etherons and their relation with mind, might help to overcome so many existing confusions. Joseph Ratzinger (Em. Pope Benedict XVI) wrote already in his book, Berührt vom Unsichtbaren,[13] 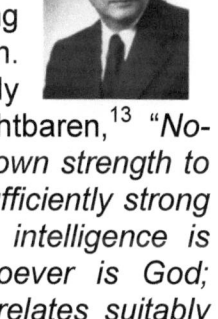 *"Nobody can build the bridge by his own strength to the infinite. No human being is sufficiently strong to call the infinite his own. No intelligence is enough to certainly devise whoever is God; whether he hears us; how one relates suitably towards him. Therefore, a particular conflict can be determined in the whole religious and intellectual history on the question of God."* Words, in an effort to understand the basics of our universe, life and mind, which is also the purpose of this document.
- Certainly some progress in science and religion still has to be made. In particular on the real cause of gravity and also on human indoctrination, with respect to matter and mind. Our hu-

---

[13] Joseph Ratzinger, Berührt vom Unsichtbaren see page 30 Decembre Jahreslesebuch, HERDER, Freiburg-Basel-Wien 2005.

man brains have sufficient capacities to solve these items, at least when these brains are not blocked from a young age on.[14]

Summary

One of the major recent discoveries concerns the super-smallest particles (or units of energy), the etherons, in our universe. They were already predicted by Albert Einstein as substrate for the transfer of electro-magnetic radiations.

With their infinitely small size of 10^{-35} m and their infinitely high density of 2×10^{36} kg/m^3 and an amount of 10^{122} in the universe, they play an essential role in its existence.

Similar as with sound: *no air molecules no sound*, there is *in absence of etherons in the cosmos no light, no telephone mobiles, no gravity and possibly no mind.*

These etherons are much too small to be detected but they nevertheless provide an explanation to a number of questions, such as:

- Have clusters of these super-high density etherons been the source of the Big Bang? Probably, because they might explode due to this high density as occurs with nuclear elements like uranium and plutonium. If so, other small Big Bangs might have occurred and can still happen. New-born stars would support this theory.

---

[14] See Reference 3, in which book (Zwikker) these items are central, and **for children** the Fairy Tale trilogy Ody (Ody his Odyssey, King Ody and the four existences and Ody's Daughter and the Solve tower), also as Amazon Kindle e-book (English) and XinXii ebook (Dutch and Russian).

- The distance between the etherons is 10^{-15} m which is the smallest wavelength of electromagnetic radiation known. Suppose that in the dark holes in the universe, radiations with smaller wave-lengths exist, these cannot escape.

Recently, gravity waves have been measured from colliding stars. These waves are produced from fluctuations in the etheron field as longitudinal compression/rarefaction waves of etheron number density.

- Mind acts in our body with as substrate the brain cells, their molecules and the etherons in the atom ether. This means that mind has an affiliation to electro-magnetic forces, otherwise it cannot steer our muscles. It acts in billions of our brain cells.

- Since mind is free within these atom ethers, mind can in principle also act outside of them. There is little difference between the atomic and cosmic ether.

- Could there be life after death? Technically speaking yes. This would confirm what mankind knows already and is preached by our various religions.

Résumé

L'une des découvertes majeures récentes est celle des super-plus petites particules (ou unités d'énergie), les éthérons dans notre univers. Albert Einstein les avait déjà prédites comme substrat pour le transfert de radiations électromagnétiques.

Avec leur taille infiniment petite de 10^{-35} m et leur densité infiniment élevée de 2×10^{36} kg/m^3 et une quantité de 10^{122} dans l'univers, ils jouent un rôle essentiel dans son existence.

Semblable au son: ***pas de molécules d'air pas de son***, on a : ***pas d'éthérons dans le cosmos pas*** *de lumière, pas de téléphone portable, pas de gravité et peut-être pas d'esprit.*

Ces étherons sont beaucoup trop petits pour être détectés, mais ils fournissent néanmoins des explications à un certain nombre de questions, telles que:

- Est-ce-que des groupes de ces étherons de très haute densité ont-elles été la source du Big Bang? Probablement, parce qu'ils pourraient exploser, comme les éléments nucléaire de l'uranium et du plutonium. Si c'est le cas, d'autres petits Big Bang pourraient avoir eu lieu et se produire encore. Les étoiles nouveau-nées soutiendront cette théorie.
- La distance entre les étherons est de 10^{-15} m, une distance égale au plus petite longueur d'onde du rayonnement électromagnetique connu. Supposons que dans les trous noirs de l'univers, des radiations de plus petites longueurs d'onde existent, celles-ci ne peuvent pas s'échapper.
- Des ondes gravitationnelles sont récemment mesurées à partir d'étoiles en collision. Ces ondes sont produites par les fluctuations dans le champ des etherons, comme des ondes longitudinales de compression/raréfaction de la densité du nombre des étherons.
- L'esprit agit dans notre corps avec les cellules du cerveau comme le substrat, leurs molécules et les éthérons dans l'éther de l'atome. Cela signifie que l'esprit a une affiliation aux forces électromagnétiques, sinon il ne peut pas diriger nos muscles. Il agit dans des milliards de nos cellules cérébrales.
- Puisque l'esprit est libre au sein de ces éthers atomiques, l'esprit peut en principe aussi agir en dehors d'eux. Il y a peu de différence entre l'éther atomique et l'éther cosmique.
- Peut-il y avoir une vie après la mort? Techniquement parlant oui. Cela confirmerait ce que l'humanité sait déjà et est prêché par nos diverses religions.

Adresses:

Acad. Prof. Ioan-Iovitz Popescu
Strada Fizicienilor 6 (new number 14), Bl. M4, Apt. 6
Magurele, Ilfov 077125
Romania
iovitzu@gmail.com

Prof. Egbert K. Duursma
302 av du semaphore
06190 Roquebrune Cap Martin
France
duursma@orange.fr.

www.ingramcontent.com/pod-product-compliance
Lightning Source LLC
Chambersburg PA
CBHW041944240526
45473CB00033B/511